That Which Is Given To Us All

IMPORTANT NOTICE

This booklet is intended to be an introductory guide, enhancing the reader's enjoyment and understanding of the flora of the coastal rainforest. I am neither a botanist nor a physician, nor am I aware of the particular physical conditions or sensitivities of any individual reader. Before attempting to use any plant as food or medicine, the reader should carefully consult additional botanical sources to confirm the identity of the plant and consult with a medical professional to ensure the use of the plant is appropriate for that reader's individual circumstances.

All plants in volume 2 grow in most or all of the coastal rainforest. Many plants listed below cross all habitats, or more than the habitat under which they are listed. In general, plants are arranged alphabetically.

Freshwater wetlands (bogs, muskegs, swamps, and wet areas):

Forest openings and open grassy meadows:

Carol: "Porcupine, why are you in a book about
          wild edible and medicinal plants?"

Porcupine: (Tlingit Name: **x̲alak' ách'**)
"Remember that day last spring when you saw me
on the side of the road as you walked up the hill?
You greeted me and said you were glad to see me,
then asked if I would visit with you for awhile.

As you waited for my response, you heard waves
lapping gently on the beach below, you felt the
warm soft breeze brush your skin, and you felt
yourself suddenly relax. You recognized by your
good feelings that I was comfortable visiting with
you.

Then you asked me if I had anything to teach you
today. Again you waited. You noticed me eating
my breakfast and asked what I was eating. I
allowed you to get very close to me so you could
see that young dandelion leaves were just
perfectly delicious right now.

Porcupine continued: "You moved too close and I let you know I felt unsafe by showing you my quills, which are quite painful if they get stuck in your skin. Fortunately, you understood my language and backed away.

After thanking me for the visit, I watched you walk up the road a ways. You found a patch of young dandelion greens and started munching!" *(Based on Project NatureConnect Activities)*

Michael J. Cohen, Ed.D. provides a unique environmental education program that has progressively helped me interact and participate in an integrated way with plants, animals, birds, fish, water, sky, earth—all of nature!

Dr. Cohen is founder and Executive Director of Project NatureConnect, established in 1987 in cooperation with the Institute of Global Education—a United Nation's Non-Governmental Organization. This program provides home study and hands-on experiences in Applied Ecopsychology and Integrated Ecology.

PNC Nature-Psychology *in action*:
Home study courses, degrees, careers
www.ecopsych.com 888.285.4694

ACKNOWLEDGMENTS

My gratitude and thanks to the Life Force that provides everything, and to everyone who supported this project:

My daughter Tiffany Hirst for her faith and confidence; my father and mother Glen and Margaret Prince, my sisters Judy Hundley and Janice Prince, and my brother Bruce Prince for their significant support and individual contributions; Phil Johnson for his support, knowledge, computer expertise, and persistence; Nancy Karacand for her unconditional support; Jeri Museth, Stan Reddekopp, and Judy Vick for their love; Norma Strickland for her lifelong friendship and business sense; Judy Jones for her friendship and editing contributions; Nora Laughlin for her friendship and generosity; Bridget Riley for her friendship and motivation.

Marie Olson of the Eagle Moiety Wooshkeetaan Clan of Auke Bay, Alaska, provided Tlingit names and usage for a number of plants in this booklet. With permission, I am including available Tlingit plant names in recognition and honor of indigenous cultures everywhere who have given the world valuable knowledge.

INTRODUCTION

On a frozen winter day last year, I spent most of my walk admiring silver-frosted, ice crystaled spiders' webs suspended in baby spruce and hemlocks. I was drawn to web after web and realized they were there all the time but could be seen only under certain conditions.

While I thought about this notion, an unusually frosty spider's web caught my attention. As I bent down for a close-up look, my focus began shifting back and forth between the web and the calm, smooth, sun-sparkling incoming tide behind the web. Junctions on the web flashed in rhythm with the tide, like a star-twinkling night sky. As I continued looking back and forth at the web and the water moving rhythmically behind the web, I sensed the invisible web of life that connects and supports all of us who inhabit planet earth.

Dandelion or violet, pebble or mountain, human or spider or spruce, made of the same basic elements, we are all related.

Common Name: **BUCKBEAN**
Other Names: Bogbean
Scientific Name: *Menyanthes trifoliata*
Plant Family: Menyanthaceae "The Buckbean Family"
Habitat: Muskegs, marshes, ponds, lake edges

Edible and/or medicinal use:

A stand of flowering buckbean gracing the edges of a marshy wetland satisfies my sense of beauty and symmetry. Buckbean's pink-tipped, feathery, white flower demonstrates nature's unparalleled artistry.

Roots can be eaten as emergency food after cooking them in at least several changes of boiling water, though buckbean is known more for its medicinal properties.

Tea made from <u>dried</u> leaves has been used as a tonic, digestive aid, diuretic (relieves water retention), laxative, an appetite stimulant, and for rheumatism and arthritis.

Common Name: **CATTAIL**
Scientific Name: *Typha latifolia*
Plant Family: Typhaceae "The Cattail Family"
Habitat: Marshes, lake edges, ponds

Edible and/or medicinal use:
This wetland plant has been used by indigenous people wherever it grows. As food, medicine, stuffing for bedding, insulation, wound dressing, and diapers, cattail offers many versatile uses. They were used during World War I to make artificial silk. Currently they are one of several plants being used experimentally to clean raw sewage.

Young female flowers, which turn to the color and texture of rich brown velvet at maturity, are edible. Rootstocks and the inner base of the young stalk are edible as well.

Before flowers appear it is possible to mistake wild iris—a potentially toxic plant—for cattail because of leaf similarity.

Common Name: **MARSH CINQUEFOIL**
Other Names: Marsh fivefinger
Scientific Name: *Comarum palustre*
Plant Family: Rosaceae "The Rose Family"
Habitat: Freshwater marshes, bogs, wet meadows,
          streambanks

Edible and/or medicinal use:

Marsh cinquefoil is in its seed stage in the picture
shown. The unique purple-brown flower might
be missed because it blends well with its marshy
environment.

A member of the rose family, all parts of the
plant may be harvested in spring and summer. It
is used for a variety of medicinal purposes. Tea
made from dried leaves may be useful in treating
diarrhea, fevers, and ulcers.

Moderate use of this plant is recommended
because of its tannic acid content, which may
irritate kidneys or stomach in large doses.

Common Name: **COTTON GRASS**
Tlingit Name: **sháchk kax̲'wáal'i**
Other Names: Alaska cotton
Scientific Name: *Eriophorum chamissonis*
Plant Family: Cyperaceae "The Sedge Family"
Habitat: Muskegs, marshes, wet ditches

Edible and/or medicinal use:
   Cotton grass, most notably used in dried flower
   arrangements, can be used as a nutritional source
   in survival situations. The base can be eaten raw,
   added to soup, or cooked with other vegetables.

   Cotton grass rootstock has properties for treating
   coughs and colds, but medical supervision is
   necessary because of possible side effects.

   Pictured is a typical freshwater marsh meadow
   scene speckled with cotton grass. A few of the
   plants one might find in this habitat include bog
   cranberries, buckbean, marsh cinquefoil, and
   marsh marigold.

Common Name: **CROWBERRY**
Scientific Name: *Empetrum nigrum*
Plant Family: Empetraceae "The Crowberry Family"
Habitat: Muskegs, alpine

Edible and/or medicinal use:
Crowberries grow abundantly in muskegs, adding their attractive evergreenness to plants that share their sphagnum mossy home, like bog cranberries, cloudberries, dwarf dogwood (more commonly called bunchberries), lingonberries, and sundew.

These berries taste pretty bland raw but improve dramatically when sweetened and cooked. As with cranberries, they are best picked after the first frost, and can be used in all baked goods, such as muffins, pancakes and breads. They may be mixed with other fruits and berries and made into jellies, juices, and syrups.

Tea made from the twigs and stems has been used for treating colds and kidney problems.

Common Name: **MARSH MARIGOLD**
Other Names: Cowslips
Scientific Name: *Caltha palustris*
Plant Family: Ranunculaceae "The Buttercup
                 Family"
Habitat: Marshes, ponds, creeks, river edges

Edible and/or medicinal use:
Large yellow marsh marigold blossoms open at
sunrise and close at sunset. When not in flower,
this plant may be confused with deer cabbage and
wild calla.

Leaves may be eaten after boiling in several water
changes to remove toxic properties. Drying them
completely will also remove toxic properties.

Medicinal tea has been helpful in relieving
cramping (antispasmodic) and promoting mucous
drainage (expectorant). Too much, however, can
be harmful to the liver and kidneys.

Common Name: **YELLOW PONDLILY**
Scientific Name: *Nuphar polysepalum*
Plant Family: Nymphaeaceae "The Water Lily
                 Family"
Habitat: Ponds, lake margins

Edible and/or medicinal use:
   Yellow pondlily is a marsh plant with long stems
   that give it the ability to grow in deep water.
   They may be rooted in the bottom or free
   floating.

   Flower seeds are edible. Reviews on edible
   properties of roots are mixed, though coastal
   Native groups used the root both as food and
   medicine.

   Medicinal preparations have been used to treat
   skin conditions, colds, tuberculosis, asthma, and
   rheumatism.

Common Name: **SWEET GALE**
Other Names: Bog myrtle, wax myrtle
Scientific Name: *Myrica gale*
Plant Family: Myricaceae "The Wax Myrtle Family"
Habitat: Bogs, fens, salt marsh edges, lake margins,
              wet coastal meadows

Edible and/or medicinal use:

Sweet gale, a deciduous bush, improves wetland soil by adding nitrogen. Leaves, which can be collected in spring and summer, are toothed only at the outer edges, a helpful identifying characteristic.

Tea, used moderately, has been recommended to treat colds, flu, and congestion. The plant contains toxic oil that evaporates easily when cooked in an open pot.

A Modern Herbal says that branches of sweet gale were used as a substitute for hops and made into gale beer. Sweet gale's fragrant leaves have been used to scent linen closets.

Common Name: **VIOLET**
Scientific Name: *Viola sp.*
Plant Family: Violaceae "The Violet Family"
Habitat: Species determines the habitat – bogs,
                moist meadows, streambanks, wet forest
                areas, subalpine
Two species are shown: (*Note the Alaska violet
identification is not verified botanically because I
did not examine the root system.)
    Stream violet: Viola glabella (yellow)
    Alaska violet: Viola langsdorffii (large purple)*

Edible and/or medicinal use:
    Violets look like they should taste sweet, so I was
    disappointed when they tasted rather acidic.
    High in vitamin C, however, acidity makes sense.
    Delicate violet flowers and leaves may be eaten
    in a variety of ways.

    Young leaves and flowers can be dried and used
    in tea to treat colds, congestion and as a mild
    laxative. Violet blossom water may be used in
    bath water to nourish and soften the skin, or as
    soothing eyewash.

Common Name: **LARGE-LEAF AVENS**
Scientific Name: *Geum macrophyllum*
Plant Family: Rosaceae "The Rose Family"
Habitat: Open forests, roadsides, meadows

Edible and/or medicinal use:
Large-leaf avens grows throughout the coastal
range and is easy to identify once you've seen its
picture.

Leaves at the base of the plant are round. The leaf
at the end of the basal stalk is much larger than the
smaller leaves below. Leaves closer to the flower
are three-lobed.

Roots and leaves have been used medicinally to
treat diarrhea, stomach pains, sore throat,
rheumatism, and as a spring tonic.

Coastal Native groups used leaves as poultices for
boils and in tea form as a diuretic.

Common Name: **SWEET COLTSFOOT**
Scientific Name: *Petasites nivalis*
Plant Family: Asteraceae "The Sunflower
                Family"
Habitat: Moist to wet meadows, along streams,
           damp places

Edible and/or medicinal use:
Sweet coltsfoot flowers before it produces leaves—an unusual characteristic. Leaves are easily recognized by their felty, white undersides.

Though flowering stalks, leaves and roots have long been considered edible, controversy currently exists. Long-term use is not recommended because coltsfoot, like comfrey, contains compounds that can be harmful to the liver.

Medicinally coltsfoot is well known by herbalists for its ability to relieve irritation of mucous tissue. It has been useful in treating lung conditions like bronchitis, whooping cough, and irritating chronic coughs. Both leaves and flower stalks are used in syrups for coughs.

Common Name: **WESTERN COLUMBINE**
Scientific Name: *Aquilegia formosa*
Plant Family: Ranunculaceae "The Buttercup
                 Family"
Habitat: Meadows, beaches, roadsides, forest
              openings

Edible and/or medicinal use:
   Brilliant red-orange Western columbine flowers
   are edible, though only moderate use is
   recommended. They can be eaten raw as a snack,
   candied as a dessert decoration, or used as a
   decorative addition to salads.

Roots and seeds are poisonous.

Gray-green columbine leaves add visually
pleasing color, pattern, and shape to floral
arrangements.

Medicinal preparations for <u>external</u> use have
been made from roots and seeds.

Common Name: **COW PARSNIP**
Tlingit Name: **yaana.eit**
Other Names: Wild celery
Scientific Name: *Heracleum maximum*
Plant Family: Apiaceae "The Carrot Family"
Habitat: Meadows, open woods, streambanks,
            beaches, roadsides

Edible and/or medicinal use:
Cow parsnip crosses most habitats, from the coast
to subalpine areas. In early spring <u>peeled</u> young
stalks and stems can be eaten either raw or boiled.
Marie Olson, Tlingit elder, recommends eating
this plant when it is one foot to 16" tall.

Some people are extremely sensitive to the outer
skin of the plant. When my brother was young, he
developed red welts on exposed skin while
weeding cow parsnip in an uncultivated area of
the yard. Skin discoloration from the welts took a
long time to disappear.

Tea from dried roots has been used medicinally in
treating nausea, colds, and heartburn.

Common Name: **TRAILING BLACK CURRANT**
Tlingit Name: **shaax̲**
Scientific Name: *Ribes laxiflorum*
Plant Family: Grossulariaceae "The Currant Family"
Habitat: Moist forest openings, meadows,
roadsides, clearings

Edible and/or medicinal use:

Botanical references classify currants in the saxifrage family or the currant family. Since volume 1 was first printed, I have noticed more references use the currant family classification, and have reflected that change here.

Trailing black currants grow along the ground, have larger leaves than bristly black currants, no bristles or stickers, and berries have a waxy looking coating. The Tlingit mashed up these currants and mixed them with salmonberries.

Medicinally, berries and leaves have been used to help people who are recovering from illness, possibly because they may stimulate the appetite. Leaves must be very <u>fresh</u>, or <u>thoroughly dried</u>, if used in tea.

Common Name: **ARCTIC EYEBRIGHT**
Scientific Name: *Euphrasia arctica*
Plant Family: Scrophulariaceae "The Figwort
                Family"
Habitat: Open, grassy, moist meadows, roadsides

Edible and/or medicinal use:
Volume 1 of this series shows a picture and gives
a description of yellow rattle, which may be used
as a substitute for eyebright.

Flowering eyebright is used as an eyewash to
treat inflammation, stinging, weeping eyes,
oversensitivity to light, and children's pinkeye.

Neither yellow rattle nor arctic eyebright is used
as food.

Common Name: **ROBERT GERANIUM**

Other Names: Herb Robert, red herb

Scientific Name: *Geranium robertianum*

Plant Family: Geraniaceae "The Geranium Family"

Habitat: Forest openings, grassy areas, clearings

Edible and/or medicinal use:

Robert geranium is an introduced Eurasian plant that is not common throughout the range of the rainforest. It is particularly attractive as the leaves turn bright red in fall, making an eye-catching contrast to its delicate pink flower. Some botanical references say this plant has an offensive odor, but I found its distinctive scent earthy and musky.

Robert geranium is used to stop bleeding.

This plant was used to treat a German plague, according to *Plants of the Pacific Northwest Coast*, by Pojar and Mackinnon.

Common Name: **MONKSHOOD**
Scientific Name: *Aconitum delphinifolium*
Plant Family: Ranunculaceae "The Buttercup
                 Family"
Habitat: Forest openings, meadows, streambanks

NO EDIBLE FOOD USE

********EXTREMELY DEADLY********

All parts of monkshood are deadly.

Monkshood can be confused with wild geranium
(shown in volume 1) before the plants flower.

Extreme caution is essential while learning about
food and medicinal use of wild plants. This point
cannot be emphasized enough!

Common Name: **SASKATOON**
Tlingit Name: **gaawák**
Other Names: Pacific serviceberry, Juneberry
Scientific Name: *Amelanchier alnifolia*
Plant Family: Rosaceae "The Rose Family"
Habitat: Forest edges, open woods, meadows,
            roadsides

Edible and/or medicinal use:

Saskatoon berries—what a treat! White flowers, gray-green leaves, and purplish-black berries of this small tree provide enjoyment at many levels. An important food source for wild birds, moose and deer, I think the birds claimed most of the berries of the two trees I had been regularly visiting.

They can be eaten raw, or prepared like other berries–in pies, jams, jellies, muffins, cereals, and pancakes. They may be stewed, dried, steamed, frozen, or blended with other fruits.

Native groups dried Saskatoon berries into hard cakes and stored them for later use.

Common Name: **SITKA VALERIAN**
Scientific Name: *Valeriana sitchensis*
Plant Family: Valerianaceae "The Valerian Family"
Habitat: Moist meadows, open subalpine

Edible and/or medicinal use:
Many species of valerian occur around the world. European valerian is considered the official medicinal species. However, Mrs. M. Grieve, in *A Modern Herbal,* states that the species pictured here is considered by some as the most powerful of all the valerians.

Valerian is generally most widely recognized for its ability to calm nerves and promote sleep. Other symptoms treated by valerian include cramps, migraine, epilepsy, and heart palpitations.

On some people valerian's sedative effect does not work. Instead it acts like a stimulant.

Common Name: **EARLY BLUEBERRY**
Tlingit Name: **kanat'á**
Scientific Name: *Vaccinium ovalifolium*
Plant Family: Ericaceae "The Heather Family"
Habitat: Species determines the habitat — woods,
             muskegs, moist meadows, alpine

Edible and/or medicinal use:

The Tlingit have five names for different
blueberry species: naanyaa kanat'aayí (big
blueberry); ts'éekáxk'w (mountain blueberry);
láx' loowú (swamp blueberry); kakatlaax
(blueberry with whitish coating); and the
blueberry pictured.

A favorite with most people, tasty and versatile
blueberries can be used in any way you can
imagine!

They may help maintain stable blood sugar,
stimulate appetite, regulate bowels, and expel
worms. Tea from leaves may be used in
moderation to treat urinary disorders.

Common Name: **FRINGE-CUPS**
Scientific Name: *Tellima grandiflora*
Plant Family: Saxifragaceae "The Saxifrage Family"
Habitat: Moist forest, forest edges, stream banks,
          shady areas

Edible and/or medicinal use:

The saxifrage genus as a whole is one of my favorites because I like the texture, shape, and spectacular reds of the leaves of some species in autumn. Within the last two years I have been attracted to a variety of species within this group.

Though I have not yet experimented with these plants as food or medicine, according to Janice Schofield, *Discovering Wild Plants*, the saxifrage genus is generally safe. Pojar and Mackinnon, *Plants of the Pacific Northwest Coast*, say one of the indigenous groups made tea from fringe-cups and used it for any ailment, particularly loss of appetite.

In the picture shown, follow the stalk down from the flowers for leaf definition.

Common Name: **SITKA MOUNTAIN ASH**
Tlingit Name: **kalchaneit**
Other Names: Rowan
Scientific Name: *Sorbus sitchensis*
Plant Family: Rosaceae "The Rose Family"
Habitat: Forests, open woods, meadow edges

Edible and/or medicinal use:

In late October I saw a group of crows land in a mountain ash and strip the tree of most of its bright red berries. Frost had sweetened the berries and attracted our black feathered friends.

Mountain ash berries picked after the first frost and mixed with apples make jellies and jams that taste especially good with wild game. High in pectin, the berries can be mixed with other berries as a jelling ingredient. Their high vitamin A and C content provides nutritional value.

Drinking a weak solution may be helpful for treating stomach and intestinal problems. Gargling with juice from the berries may relieve sore throat.

Common Name: **OCEAN SPRAY**
Other Names: Creambush
Scientific Name: *Holodiscus discolor*
Plant Family: Rosaceae "The Rose Family"
Habitat: Open woods, clearings, coastal bluffs

Edible and/or medicinal use:

Sweet-scented ocean spray is a Northwest native shrub with creamy white flowers that change in June and July from white to cream and to reddish tan as the seeds mature in early fall.

Ironwood, another common name for this bush, was used by Native people. Branches were hardened over fires and shaped into a variety of tools.

According to *Plants of the Pacific Northwest Coast*, by Pojar and Mackinnon, brownish fruiting clusters of ocean spray were steeped in boiling water and made into a drink for treating diarrhea. The plant was used to treat measles and chickenpox and was known as a tonic for the blood.

Common Name: **OLD MAN'S BEARD**
Other Names: Common witch's hair,
                    Methuselah's beard
Scientific Name: *Usnea longissima*
Plant Family: Parmeliaceae "The Parmilia Family"
Habitat: Trees, shrubs

Edible and/or medicinal use:

Currently, I am researching Usnea because of its antibiotic, anti-fungal, anti-viral, and anti-bacterial properties. It fights bacteria and viruses that contribute to bronchitis, pneumonia and other lung problems. Conditions such as Candida overgrowth, urinary tract problems, and painful inflammations have been successfully treated with Usnea. We will be hearing lots more about this important lichen.

Old man's beard was called the "lungs of the earth" by some indigenous cultures. It hangs in long strands on trees and bushes and depends on rainwater for nutrients. Its presence indicates clean air because of its extreme sensitivity to pollution.

Common Name: **SALAL**
Scientific Name: *Gaultheria shallon*
Plant Family: Ericaceae "The Heather Family"
Habitat: Coniferous forests, open areas

Edible and/or medicinal use:
   Covering more than half the range of the
   rainforest, salal is one of the most common forest
   shrubs.

   Berries can be used to make jam, jelly and syrup.
   They can be added to baked goods such as
   muffins, cakes, and pies, and to oatmeal and
   other cooked cereals, or blended into yogurt and
   fruit smoothies.

   An important fruit for all Northwest coast Native
   people, the berries were eaten fresh, as well as
   dried for winter use.

Common Name: **BEACH ASPARAGUS**
Other Names: American glasswort, pickle weed
Scientific Name: *Salicornia virginica*
Plant Family: Chenopodiaceae "The Goosefoot
                 Family"
Habitat: Salt marshes, tide flats

Edible and/or medicinal use:

American glasswort is a more commonly used name for this plant. I prefer common names that relate to something I know, and beach asparagus, though smaller, looks like asparagus one buys in the grocery store. It has also been called pickle weed or pickle plant because it makes excellent pickles, which I plan to make next spring or summer. Janice Schofield provides a recipe in *Discovering Wild Plants*.

The common name glasswort derives from the use of its ashes in making glass commercially.

Beach asparagus tastes great steamed. Juice from the fresh plant has been used as a diuretic.

Common Name: **CLEAVERS**
Scientific Name: *Galium aparine*
Plant Family: Rubiaceae "The Madder Family"
Habitat: Beaches, forest edges, open areas, ditches

Edible and/or medicinal use:
By means of tiny hooked bristles, cleavers stick to
other plants and passing objects, allowing this weak-
stemmed herb to establish itself among sturdier plants
and move upwards into sunlight.

Cleaver shoots collected in early spring can be eaten
as a steamed vegetable.

Used as a folk remedy for centuries, cleavers has been
helpful for many types of skin problems, as a diuretic,
blood purifier, lymphatic cleanser, and tonic.

It is not recommended for people with diabetic
tendencies.

Tea from the dried plant has been used as a treatment
for insomnia.

Common Name: **OYSTERLEAF**
Other Names: Sea bluebells
Scientific Name: *Mertensia maritima*
Plant Family: Boraginaceae "The Borage Family"
Habitat: Coastal areas, sandy beaches

Edible and/or medicinal use:

When I take walks along coastal sand beaches in my area, gray-green oysterleaf especially catches my eye. Smooth and somewhat succulent looking, the common name stems from its oyster-like tasting leaves. From spring to summer they may be added to soup, stew, salad, and egg dishes.

New leaves continue growing until freezing weather, offering a tasty snack on cold autumn beach walks.

Lovely pink buds turn into bell-like blue flowers that may be eaten as well.

Oysterleaf offers an attractive contrast to some of the surrounding greens.

Common Name: **SILVERWEED**
Scientific Name: *Potentilla anserina*
Plant Family: Rosaceae "The Rose Family"
Habitat: Seashores, upper beaches, tidal meadows

Edible and/or medicinal use:
Easily recognizable, silverweed is a low-growing
plant with runners like strawberries, and lovely
bright gold flowers.

Edible silverweed roots may be steamed or boiled.

Related to marsh cinquefoil, another *Potentilla*
species found earlier in this volume, the entire
Potentilla genus is considered medicinal.

Flowers and stems are used medicinally in
tinctures and infusions for respiratory problems,
as an anti-inflammatory and as a diuretic.

Common Name: **RED CLOVER**
Scientific Name: *Trifolium pratense*
Plant Family: Fabaceae "The Pea Family"
Habitat: Fields, waste areas, roadsides

Edible and/or medicinal use:
I remember sucking the juice from red clover
when I was a kid, always wishing for more nectar
in each of the individual pieces of the flower.
Some were so sweet they felt like a prize.

When I drive past clover fields in the summer
with my car windows down, I am delighted by its
sweet scent.

All of the plant is edible. Leaves, stems and
flowers may be added to many foods—salads,
soups, stews. Clover honey is delicious,
especially on biscuits hot from the oven.

Medicinally, red clover is considered an
exceptional blood purifier. Tea from flowers may
be helpful in treating hepatitis, mononucleosis,
headaches, arthritis, coughs and nerves.

Common Name: **COMMON DANDELION**
Tlingit Name: **yeil <u>k</u>a<u>k</u>wu**
        (translates as Raven's cup in English)
Scientific Name: *Taraxacum sp.*
Plant Family: Asteraceae "The Sunflower Family"
Habitat: Roadsides, meadows, disturbed areas

Edible and/or medicinal use:

All parts of the dandelion are edible at various times of the year—new leaves in spring, flowers in summer, roots in fall. They may be dried, steamed, stir-fried, or added to soups, casseroles, omelettes, and stews.

Well known by herbalists as the liver herb, dandelions may be useful in lowering cholesterol, high blood pressure, and stabilizing emotions.

They have been used to treat liver problems, gastritis, kidney disease, gall bladder inflammation and dyspepsia. Dandelion root teas are recommended for cases of hepatitis, anemia, jaundice, cirrhosis and gallstones, as well as edema, gout and rheumatism.

Common Name: **DOCK**
Scientific Name: *Rumex sp.*
Plant Family: Polygonaceae "The Buckwheat
                Family"
Habitat: Wet, disturbed soil, roadside ditches

Edible and/or medicinal use:
Dock would be a good plant to learn in fall when it has gone to seed. A large, showy plant, it can produce spectacular reds in its leaves and seeds.

Young dock leaves may be used in moderation in salad, soup and stew. They taste sour and lemony when they are eaten raw. <u>Older leaves must not</u> be eaten because they contain oxalic acid, which can be toxic.

Yellow dock is the species used for medicinal purposes. It has been used to treat anemia, hepatitis or liver damage, constipation, blood disorders, skin diseases, rheumatism and indigestion.

Common Name: **GOATSBEARD**
Scientific Name: *Aruncus sylvester*
Plant Family: Rosaceae "The Rose Family"
Habitat: Roadsides, forest edges, meadows

Edible and/or medicinal use:
A showy member of the rose family, goatsbeard
ranges throughout the rainforest.

Thus far, my personal experience and knowledge
of goatsbeard is limited to my appreciation of its
lush fluffy white bloom in summer, and
spectacular orange, red, and gold displays in
autumn. On my almost daily walks, I sense
companionable gentleness from goatsbeard.

According to Pojar and Mackinnon, the Tlingit
used the root to treat blood diseases, as did other
Native people along the coast. In addition coastal
indigenous groups used roots as a diuretic and
cough medicine, and to treat swellings and kidney
pain.

Common Name: **PEARLY EVERLASTING**
Scientific Name: *Anaphalis margaritacea*
Plant Family: Asteraceae "The Sunflower Family"
Habitat: Roadsides, open forest, waste places,
        meadows

Edible and/or medicinal use:
*The experience of physical, mental, emotional and
spiritual well being decribes my definition of
medicinal:* Pearly everlasting provides emotional
and mental well being through its delicate papery
whiteness in dried flower arrangements. On dark
winter days I'm cheered by the memory of sunny
summer warmth when I collected them.

Medicinally this plant has been used in steam
baths to treat rheumatism.

Common Name: **COMMON PLANTAIN**
Scientific Name: *Plantago major*
Plant Family: Plantaginaceae "The Plantain Family"
Habitat: Roadsides, waste areas, disturbed sites

Edible and/or medicinal use:
Young plantain leaves add many nutritious benefits to spring salads. They may be lightly steamed by themselves, or mixed with other vegetables.

Plantain poultices have been used to treat infections. The Tlingit used it to treat burns. Tea or juice may be helpful in treating eczema and other skin problems, as well as kidney problems, asthma, and earaches.

Plantain is considered one of the major herbal healers, especially for a variety of infections, and blood disorders. Plantain seeds, known as psyllium, are a major ingredient in some commercial laxatives and fiber increasing products.

Common Name: **SELF-HEAL**
Other Names: Heal-All
Scientific Name: *Prunella vulgaris*
Plant Family: Lamiaceae "The Mint Family"
Habitat: Roadsides, forest edges, disturbed sites

Edible and/or medicinal use:

A little community of low purple cone-like plants growing along the roadside caught my attention a few years ago. I added their picture to my collection of unknown plants for future identification, and continued enjoying their rich purple addition to my walks while I got to know them better.

Self-heal, or heal-all, comes from Eurasia and has made itself at home all over the country.

Medicinally it has been used to stop bleeding and for many other ailments, including sore throat, colic, gas, diarrhea and thyroid imbalances.

Tea from this plant tastes mildly apple-like and very refreshing.

References

NOTE: In this small space I am limited to only a few of the many botanical and nature references used over the years.

Buhner, Stephen Harrod. (1996). <u>Sacred Plant Medicine</u>. Boulder, Colorado: Roberts Rinehart Publishers.

Cohen, Michael J. (1997). <u>Reconnecting With Nature</u>. Corvallis, Oregon: Ecopress.

Grieve, Mrs. M. (1971). <u>A Modern Herbal</u>. Vols. 1 & 2. New York: Dover Publications.

Hall, Judy Kathryn. (1995). <u>Native Plants of Southeast Alaska</u>. Juneau, Alaska: Windy Ridge Publishing.

Ody, Penelope. (1993). <u>The Complete Medicinal Herbal</u>. New York: DK Publishing, Inc.

Pojar, Jim and Andy MacKinnon, eds. (1994). <u>Plants of the Pacific Northwest Coast, Washington, Oregon, British Columbia & Alaska</u>. Vancouver: Lone Pine Publishing.

Schofield, Janice J. (1989). <u>Discovering Wild Plants, Alaska, Western Canada, The Northwest</u>. Anchorage: Alaska Northwest Books.

Interviews:

Olson, Marie. (1997 & 1998). Personal Interviews. <u>Conversations with Tlingit elder of The Alaska Native Sisterhood, Camp 2</u>. Juneau, Alaska.

Telephone and Written Interviews:

Stensvold, Mary. (1998). Personal Interviews and Written Communications. <u>Conversations with Mary Stensvold (Regional Botanist, USDA Forest Service, Alaska Region, Sitka, Alaska) verifying scientific and plant family names and species identification</u>.

INDEX